AF294984

Matthias Mueller is enterprise and sustainability consultant. He lives in Switzerland. He was president of the Swiss office of The Natural Step network and initiated and implemented the smartphone app "Sustainability Compass".

Sustainability Compass.

Think smarter, act smarter

A guide for sustainable decision-making

Matthias Mueller

emem publishers / emem.ch
compass-for-sustainaibility.net

The questions of the Sustainability Compass gently help us realize that we WANT to live more sustainably.

Bob Willard

Your Sustainability Compass

Every day we take decisions which impact nature and human beings and their society. We decide on small and big purchases, holiday plans, business projects, plans with friends, the family or at school.

And we want to make sure that our impact is not negative for men and nature. We want to protect what nourishes us and allows us to exist.

The Sustainability Compass doesn't take away the decision-making from you. It helps you to think of and investigate smarter action.

This is how you could proceed and build a Sustainability Compass for a project: When preparing a decision, you study the twelve criteria of the Sustainability Compass. While reading and discussing them with friends, family, colleagues you will find out the four most important criteria you'd like to meet with your project. You plan your action based on these four criteria you want to respect. This is your Sustainability Compass (like the "real" compass with its four directions).

But your work isn't done yet. For every criterion there is a check list with questions which guide you into researching what the smart action would look like. This check list might create further questions and it might make you dive into more information. It will be an exciting journey to make you think smarter and act smarter. Get informed and involve other people!

What attitude is needed to get the best out of the Sustainability Compass:

- You are the champion - you build your Compass
- There must be a decision which is important and creates impact
- You are curious and willing to find valuable information - and you want to discuss it
- You are creative and never get tired finding out the options and the potential of your actions
- You are courageous and you act smarter.

After the description of the criterion you find an empty page which allows you to write notes or ideas.

In case you'd like to make others aware of the Sustainability Compass make them this booklet a gift - or download the pdf. from the website: compass-for-sustainability.net.

Image: Shutterstock

Heavy metals

There are two reasons to restrict the use of heavy metals and rare earth elements (in consumer electronics, pigments, PVC, cosmetics, etc.):

First, they contaminate people and the ecosystem – for example lead finds its way into drinking water.

Second, since they are rare their price is bound to increase – for example cobalt in mobile phones or iridium in flat screen television.

It is important to decrease our use of heavy metals and rare earth elements.

Heavy metals: Check list

- Am I aware of the presence or content of heavy metals and rare earth elements: are they clearly labelled?
- Are there alternatives which use less or no heavy metals and rare earths?
- Do I know where I can return equipment containing heavy metals where they will be professionally reprocessed?

My notes

Image: Aliencow, Rio +20 demonstration huge globe

Participation

15

Participation means to be part of and to take part. Through participation one can understand and influence what others want, and also take responsibility. Solutions in which many participate are often more robust than those developed by individuals or small groups.

It is important that people have a say about the direction their life takes.

Participation: Check list

- What is the level of stakeholder participation during the entire lifecycle?
- Are groups or individuals excluded from giving input or making decisions which affect them?
- Are the relevant actors treated with respect in an inclusive environment of trust?

My notes

Image: Shutterstock

Suitable satisfiers

Satisfiers are objects or actions which satisfy basic human needs like protection, freedom or affection. But some satisfiers tend to destroy, to block the satisfaction of other basic needs or are simply unhealthy: Excessive gambling makes us lonely and poor. A six hours flight to a wellness weekend causes stress and harms the climate.

It is important to distinguish between a need and a satisfier, and to avoid false satisfiers.

Suitable satisfiers: Check list

- Will my intended result actually satisfy me, my family and my community?
- Can I ensure that the proposed action does not block or render impossible the satisfaction of other important needs?
- Am I able to explain to a friend what human need I want to meet with the proposed action?

My notes

Image: Torsten Henning, Warnzeichen D-W003 nach DIN 4844-2

Chemicals

A wide range of man-made chemicals and plastics do not break down naturally and, once released into nature, are very difficult to remove. They do considerable damage when they get into human and animal food chains (e.g. PET in fish), water or when they contaminate soils. These materials should be collected and recycled or replaced by materials that are biodegradable.

It is important that no persistent chemicals or plastic get into nature.

Chemicals: Check list

- Am I aware of the use of chemicals/plastics during the entire lifecycle (raw materials, production, distribution, use, end of life)?
- Do I know whether and how the plastics I use can be recycled?
- Do I know whether chemicals are released into the environment (evaporation, abrasion, dissolution in water)?

My notes

Image: Shutterstock

Protection of nature

27

Nature is destroyed as a result of overfishing, deforestation, uncontrolled waste dumping or through excessive overbuilding. These activities impair nature's ability to provide the "eco-services" such as growth (plants), cleaning (water) or storage (water and energy).

It is important that nature is not destroyed by physical impact.

Protection of nature: Check list

- Do I know services and products which reduce the destruction of nature (deforestation, building on precious soil, species extinction, overfishing, pollution of soil/water/air)?
- How can my behavior help to reduce the physical destruction of nature?
- How can I prevent the erosion of topsoil? It takes nature 250 to 300 years to restore the loss of 5mm topsoil.

My notes

Image: Shutterstock

Fossil fuels

Fossil fuels provide society with inexpensive energy, but the systematic burning of coal, petroleum and natural gas has become the biggest threat to the biosphere. Greenhouse gases accumulating in the atmosphere, such as CO2 and methane, have led to massive changes in the climate and eco-systems of our planet. We must promote the use of renewable energy on a large scale.

It is important to stop burning non-renewable fossil fuels.

Fossil fuels: Check list

32

- Can I achieve my goal in a completely different way - without using any fuel?
- Is there an alternative which reduces dependence on fossil fuels?
- Can I purchase an offset for the CO2 emissions I cause - with a provider I can trust?

My notes

Image: Shutterstock

Connectedness

35

For many it is simply good to be in nature. Experiencing fresh air, sun, wind and rain gives a sense of orientation and newfound strength. Connection with nature is important to assess and learn to recognize the fragility of the eco-system.

It is important to feel connected with nature.

Connectedness: Check list

- Will the outcome lead to an increased sense of interdependence with nature or provide an opportunity to experience the beauty of nature?
- The sense of support and belonging – such as that which nature provides – fosters emotional well-being. Have I considered this?
- Does the proposed action help me to deal more carefully even in my professional environment?
- Does nature need the proposed action?

My notes

Image: Howard Dickins, Riverside Farmers Market, Riverside, Cardiff, Wales.

Fairness

Fairness means do unto others as you would have them do unto you. The price of fairness can consist of abstaining from benefits in favor of others.

As participant in economic life you can ask for and live up to the principles of fairness and transparency: sometimes it may be more expensive, but you take responsibility towards the community you live in.

It is important that we live and understand fairness in a global sense.

Fairness: Check list

- Do I use my influence to promote consumption of fairly produced products?
- Am I sure that the rights I enjoy are also available to others?
- How do I explain fairness to family, friends, acquaintances and work colleagues?

My notes

Image: Shutterstock

Health

43

Our physical and emotional health is challenged daily by deficiency, abundance or lack of balance (rest and stress). Moreover, what is healthy for one can be unhealthy for the other.

Example: Clothes protect its wearer but can be produced under conditions hazardous to health.

It is important that people try to live healthy and support the health of others.

Health: Check list

44

- Will the outcome contribute to my own health without risking the health of others?
- In what way is the health of other contributors affected?
- Am I aware of the reasons for poor health and how to avoid them?

My notes

Image: Shutterstock

Idleness

Idleness is not a vice. Researchers have found that it is one of the basic human needs. Idleness means to relax, to dream, to play, to maintain good mood and calmness. Often little is needed to satisfy this need. The get-together with friends can be a place of idleness, the deck chair, talking about the old days or a spontaneous party.

It is important to appreciate moments and places of idleness.

Idleness: Check list

Will the outcome permit those involved to enjoy times of refueling, daydreaming, doing nothing?

Is it necessary to harm nature while enjoying a state of idleness?

What about play? Can it be used to motivate responsible action?

My notes

Image: Shutterstock

Identity

51

Identity means to develop a sense of belonging and self-assurance. Lack of self-esteem often leads to unsustainable, destructive actions. Identity can manifest itself in language, religion, sexuality, group membership, career, ethical and aesthetic values.

It is important that people can live and express their identity.

Identity: Check list

52

- Will the outcome make me proud?
- Are those involved able to freely express their individuality?
- If I were to explain to someone why sustainability is important to me, what would I say?
- Does the proposed action foster or limit diversity?

My notes

Image: Shutterstock

Understanding

People want to understand - they perceive and ask questions. To get answers they ask the elderly, read books, research, make hypotheses and experiment. For many people the beauty and richness of nature intensifies when they start trying to understand it.

It is important for people to understand – themselves and the world they live in.

Understanding: Check list

56

- Will the outcome permit those involved to better understand - find something out - discover?
- Why is it that I'm curious?
- How can others benefit from my curiosity?

My notes

Scientific background of the Sustainability Compass

The Sustainability Compass is built on two frameworks.

1) The framework of the organization "The Natural Step" builds on scientific research. It defines four system conditions which have to be respected by a sustainable society.

Source: https://thenaturalstep.org

2) The Chilean economist Manfred Max-Neef wrote "Human Scale Development". While researching poverty he discovered nine basic human needs which every human being wants to satisfy - independent of sex, status, income, ethnicity or culture.

Source: Manfred Max-Neef: "Human Scale Development". The Apex Press, New York and London, 1991.

Development of the Sustainability Compass

The Sustainability Compass was initially designed as smartphone app - it still can be downloaded as such. Goal of the compass is to offer an instrument/language which facilitates smarter decisions. The development of the app consisted of several iterations with role play and expert feedbacks. The intent was to find an everyday language used by everybody which accepts scientific knowledge.

25 sustainability experts from Switzerland, Sweden, Mexico, Brazil, Canada, USA and India engaged during the development of the Sustainability Compass.

Special thanks go to: Itzel Orozco, Telma Gomes, Herman Gyr, Lisa Friedman, David Hasler, Stanley Nyoni, Richard Chrenko, Peter Carson, Reed Evans, Kathrin Fuchs and Alexandre Magnin.

Feedbacks

"Modern science has shown that the world we live in is unsustainable. The problem is, how do we reach out to create communities of people who are willing to do something about it? The Sustainability Compass app does it by asking smart questions which inspire for smarter actions. I appreciate its directness and simplicity."

Karl-Henrik Robèrt, founder of The Natural Step,
Ashoka Fellow and recipient of the Blue Planet Prize

"The packaging of the app and the card deck is really smart. The questions gently help us realize that we WANT to live more sustainably - that it is important to us."

Bob Willard, author of "Sustainability Advantage"

"In my opinion, the majority of good decisions and innovations should begin by addressing a problem or well-asked question. I like the question-asking function of the Sustainability Compass."

Andreas Gyr, Architect, Programm Manager Google

"The questions of the Sustainability Compass make you think out of your box. I like that. A good investment."

Fatima Vidal, author, blogger, artist